Sunlight, Skyscrapers, and Soda-pop

Sunlight, Skyscrapers, and Soda-pop

The Wherever-You-Look Science Book

Andrea T. Bennett
James H. Kessler
Illustrated by Melody Sarecky

McGraw-Hill

A Division of The McGraw-Hill Companies

Library of Congress Cataloging-in-Publication Data

Bennett, Andrea T.
Sunlight, skyscrapers, & soda pop: the wherever-you-look science book / Andrea Bennett, James Kessler.
p. cm.
Summary: Introduces scientific discoveries through two imaginary creatures exploring the natural world around them. Includes activities.
ISBN 0-07-001440-X (pbk.)
1. Science—Juvenile literature. 2. Science—Study and teaching (Elementary)—Activity programs—Juvenile literature.
[1. Science.] I. Kessler, James H. II. Title.
Q163.B495 1997
507'.8—dc21 97-23687
CIP
AC

Published by The McGraw-Hill Companies, Inc.

The activities described in this book are intended for children under the direct supervision of adults. Neither the American Chemical Society nor the publisher can be held responsible for any accidents or injuries that might result from conducting the activities without proper supervision, from not specifically following directions, or from ignoring the cautions contained in the text.

pbk 1 2 3 4 5 6 7 8 9 0 KGP/KGB 9 0 2 1 0 9 8 7

Acquisitions editor: Judith Terrill-Breuer
Production supervisor: Claire B. Stanley
Editing supervisor: Patricia V. Amoroso

A Word to Parents and Teachers

Welcome to *Sunlight, Skyscrapers, and Soda-pop: The Wherever-You-Look Science Book*, second in a series of hands-on science activity books for young children. As with our first book, *Apples, Bubbles, and Crystals: Your Science ABCs*, the activities in this book are designed to be done by children with an adult partner. In addition to the hands-on science activities, this book also contains a unique interactive learning feature.

How To Use This Book

Start by enjoying the rhymes that will take you through a fun science-filled day with Sally and Sammy. Whether they're in the kitchen making breakfast or playing in the park, Sally and Sammy discover that science is all around them and that science helps to explain why things happen the way they do.

For each place Sally and Sammy visit, there are two hands-on science activities. These activities relate to something Sally and Sammy are doing or to something they observe. Then, contained *within* each hands-on activity is a further *science search challenge*. Science search challenges ask the child to locate another example (somewhere in the book) of the science introduced through the activity. To help young readers locate these examples, one of Sally and Sammy's little twin siblings will appear next to each science search answer. An answer key for these search challenges as well as a brief explanation of each science activity can be found in the back of the book.

Have fun with this book. Let science come alive through the activities. And good luck in the science search!

Sunlight, Skyscrapers, and Soda-pop

Meet Sally and Sammy—
They love to have fun.
They do things together
Until the day's done.

Our Sally loves science.
She knows a whole lot.
But Sammy's still learning—
He's only a tot.

With Sally and Sammy
As they start to play,
We'll do the fun science
That's part of each day.

I didn't turn on the
Bright light that you see.
It bounced off the mirror
So please don't blame me!
It's time to get up now
To start a new day.
Yikes, look at my hairdo!
It's flying away!
Learn more about Sally's
fly-away hair and the light
in Sammy's eyes by turning
the page and doing some
fun science activities!
TURN

STATIC MAN

This is what you need:

- Styrofoam ball, about 1½" diameter
- ruler
- thread
- scissors
- pencil, sharpened
- marker
- balloon

Here's what you do:

1. Cut about 10 strands of thread, each about 2" long.

2. Use the pencil point to poke a hole in the styrofoam ball. Now wrap the bundle of threads around the pencil point and use the pencil to stick the thread into the hole.

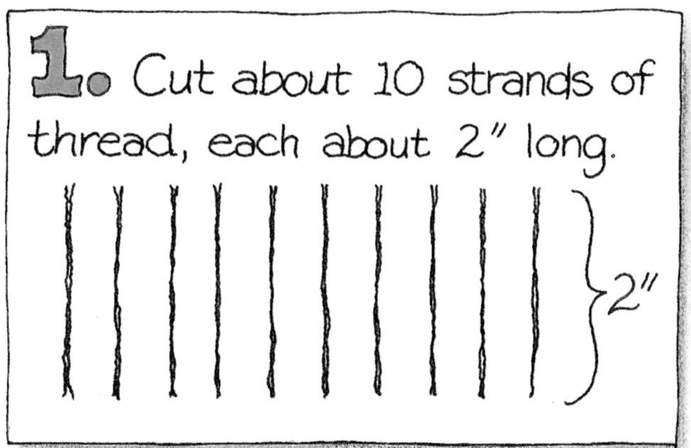

3. Now poke the pointed end of the pencil into the ball on the side opposite the thread.
Using your marker, draw a happy face.

4. Have your partner carefully blow up the balloon and tie it.
Rub the balloon several times against your hair, shirt, pants, or carpet.

Did you know that lightning is also caused by static electricity? See if you can find another example of static electricity in this book!

5. Hold the balloon next to Static Man's hair. What happens to his hair? Move the balloon around. What makes Static Man's hair move?

Bouncing Beams

This is what you need:

- flashlight
- small mirror

When light from one direction hits a smooth, shiny surface, it bounces or "reflects" off that surface in another direction.

Here's what you do:

1. Turn the lights off so that the room is as dark as possible.

2. Hold the flashlight in one hand and the mirror in the other.

3. Have your partner call out the name of an object in the room.
DOORKNOB!
4. Shine your flashlight on the mirror and then turn the mirror so the light bounces off and hits the object. Now have your partner name another object in the room and try it again.
See if you can find another example of bouncing light beams somewhere in this book!
hmmm...
tap tap

he next stop's the kitchen
For something to eat,
Where Sally finds science
All over her feet!

I'll need to add eggs to
My list for the store.
I'll add magnets, too, 'cause
We always need more!
YUMMY O's
MILK
GROCERY LIST
T A B C S
ONWARD
Do the fun activities
on the next two pages
and discover just how
strong eggs can
be and other places
where magnets stick!

The Mighty Dome

This is what you need:

- 4 eggshell halves*
- square piece of stiff cardboard, 20 cm x 20cm (8 inches x 8 inches, approx.)
- magazines or thin books
- paper towel

*NOTE: Clean eggshell halves thoroughly with soap and water before using.

Sometimes the shape of an object will make it stronger. Check out this activity to see just how strong an eggshell can be!

Here's what you do:

1. Place the egg shell halves on the paper towel in a square.

2. Lay the cardboard square on top of the egg shells.

3. Start placing magazines or thin books one by one on top of the cardboard. How many do you think you can stack before the egg shell will break?
4. Keep stacking until you hear the domes cracking. Were the domes stronger than you thought they would be?
CRACK
A dome is a shape that looks like an upside-down bowl. See if you can find another dome somewhere in this book.

Magnet Mystery Map

This is what you need:

- strong refrigerator door magnet
- crayons or colored markers

Here's what you do:

1. Have your partner make a photo copy of the kitchen map on the next page.

2. Take your magnet and your kitchen map into the kitchen. Use your magnet to test the objects listed to see which ones are magnetic.

Magnets stick to some things, but not to others. See if you can find another fun use for magnets somewhere in this book!

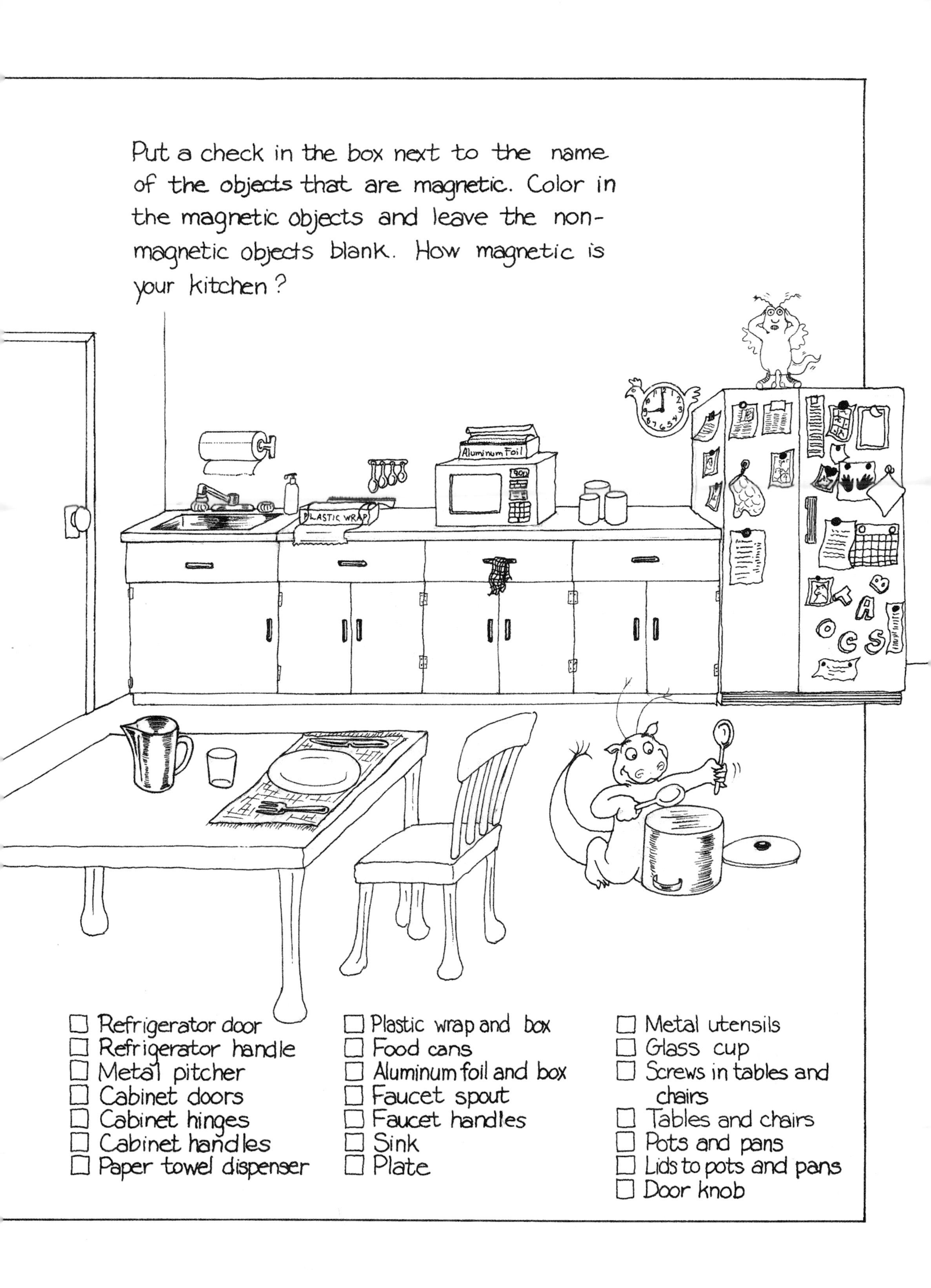
Put a check in the box next to the name of the objects that are magnetic. Color in the magnetic objects and leave the non-magnetic objects blank. How magnetic is your kitchen?
Aluminum Foil
PLASTIC WRAP
☐ Refrigerator door
☐ Refrigerator handle
☐ Metal pitcher
☐ Cabinet doors
☐ Cabinet hinges
☐ Cabinet handles
☐ Paper towel dispenser
☐ Plastic wrap and box
☐ Food cans
☐ Aluminum foil and box
☐ Faucet spout
☐ Faucet handles
☐ Sink
☐ Plate
☐ Metal utensils
☐ Glass cup
☐ Screws in tables and chairs
☐ Tables and chairs
☐ Pots and pans
☐ Lids to pots and pans
☐ Door knob

While Sammy and Sally Are walking their pet, They'll see some more science. On that you can bet!
Hey, look at that building— It's really so tall! The builders used science, So it will not fall!
Just look at this bridge, Sal. It must be quite strong. It even stays up with Our Ernie along!
The shape that they built it Helps hold up more weight. If they built it wrong, then We'll end up fish bait!

Try the next two activities
and discover how to make
bridges stronger and
buildings taller !

BRIDGE BUILDING

This is what you need:

- sheets of paper cut in half lengthwise
- two stacks of books of equal height (about 10 cm)
- blunt tip scissors
- metric ruler
- pennies

Here's what you do:

1. Place the two stacks of books on a table at least 12 cm apart.

2. Using the half sheets of paper, make a bridge across the two stacks of books. Try out each of the design ideas on the next page first and then come up with a design of your own.

The Flat Bridge

Test the strength of each bridge design by stacking pennies one at a time in the middle of the bridge. See how many pennies your bridge can support without caving in.

The One-Fold Bridge

Fold your piece of paper lengthwise and lay it across the books.

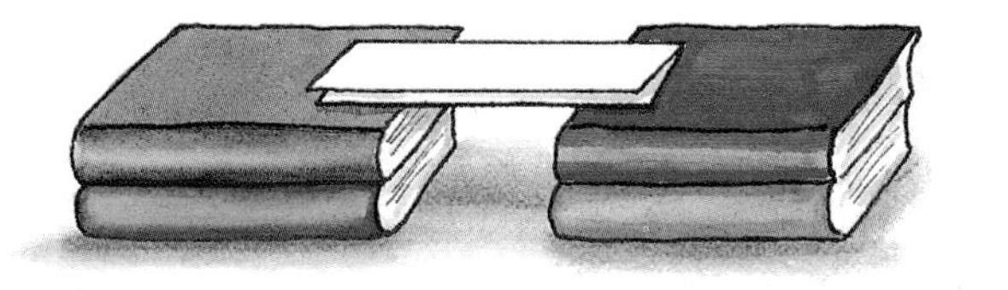

The Accordian Bridge

Fold the long side of the paper back and forth to make pleats.

The Arched Bridge

Cut the paper in half lengthwise. Place one strip between the books to make and arch and lay the second strip across the arch to make a bridge.

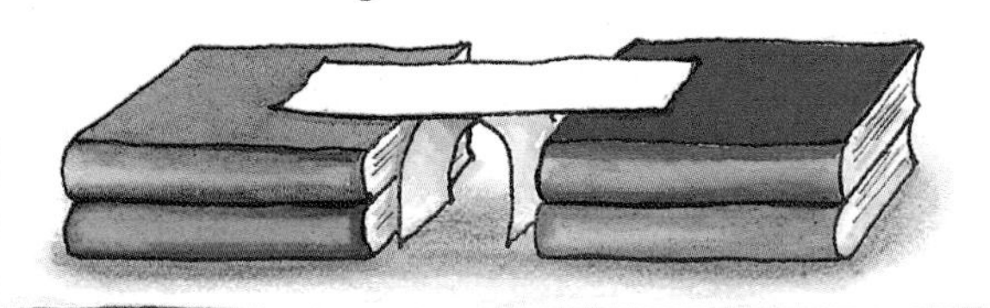

The Walled Bridge

Fold each side of the paper up so that your bridge has walls on both long sides.

Which design made the strongest bridge? See if you can find another bridge somewhere in this book.

TEETERING TOWERS!

This is what you need:

- 7-oz. plastic or paper cups
- white liquid glue
- 40-50 pennies
- magazine

Here's what you do:

1. Use a very thin layer of glue on the rim and bottom* of each cup to build a skyscraper that is 4 cups tall. Let the glue dry.

2. Wave the magazine up and down to make a windstorm.

Skyscrapers are so tall and skinny-what do you think keeps them from blowing over, in a windstorm?

I don't know. Let's do this activity and learn more about skyscrapers.

3. If your skyscraper is still standing, glue some more cups on to it to make it even taller.

4. Test your tower again with the windstorm. How tall a skyscraper can you build that will not fall over in a windstorm?

5. Now fill a cup with 40-50 pennies. Glue it to the bottom of your skyscraper. Does this help the skyscraper stand up through a windstorm? See how high you can build your skyscraper using this heavy base.

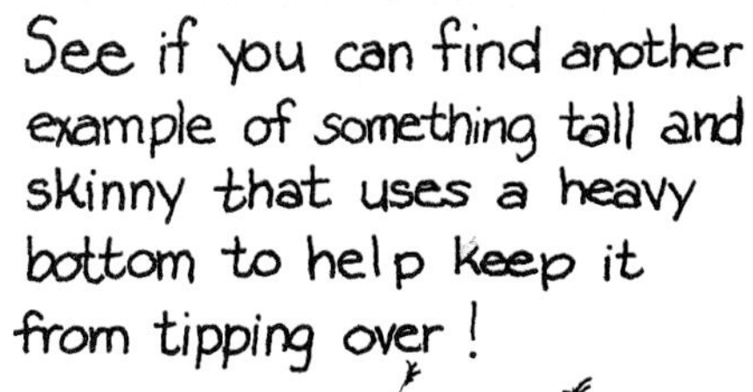

See if you can find another example of something tall and skinny that uses a heavy bottom to help keep it from tipping over!

he park is the place where
Our friends jump and run.
They play lots of games and
It's really such fun!

Be sure that your weight is
In just the right place
To keep you from falling
Down flat on your face.
And when you are finished
Hopscotching about,
Let's go do some boating.
It's fun, there's no doubt!
Explore how we balance and a new way to make boats go!

Balancing Beastie

This is what you need:

- tracing paper
- card stock or manilla folder
- blunt-tip scissors
- paper clips

We need balance for so many things we do-sitting, walking, riding a bicycle. Try this fun activity to learn more about balancing!

Here's what you do:

1. Using your tracing paper, trace Balancing Beastie to make a pattern.

2. Cut out your pattern of Balancing Beastie and trace his outline onto the card stock or manilla folder. Cut Balancing Beastie out of the card stock.

3. Place a paper clip on each of Balancing Beastie's hands as shown. Place his head on your finger tip and see if he balances.

Most people talk about hearing, seeing, tasting, smelling, and touching as the five senses. But balance is another important sense. See if you can find another example of balance somewhere in this book!

Surface Sailing

This is what you need:

- index card
- blunt-tip scissors
- baking plan
- liquid dish detergent
- water

The surface of water is stronger than it looks. It has a special kind of "skin" that can keep some objects, like this boat, from sinking!

Here's what you do:

1. Using your index card, cut out a little boat like the one above.

2. Fill the baking pan half-full with water.

Did you know that water's special skin lets some insects walk around on the surface of water? See if you can find these special insects called "water striders" somewhere in this book!

* Remember to rinse the pan well after each race or your boat won't go!

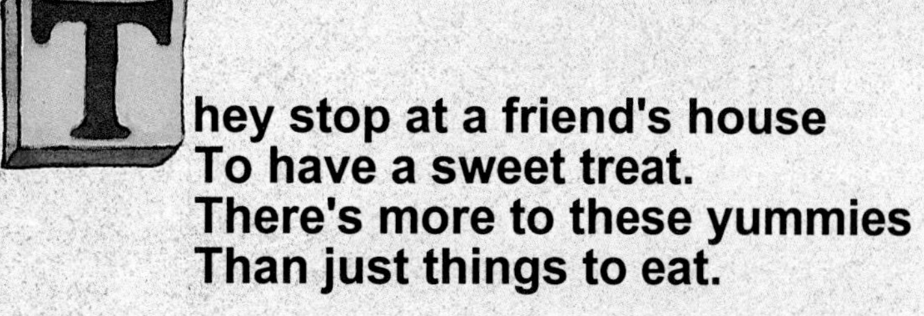

They stop at a friend's house
To have a sweet treat.
There's more to these yummies
Than just things to eat.

Discover why Sally puts
raisins in her soda and
what makes colored candy
so colorful!

RACIN' RAISINS

This is what you need:

- raisins
- clear soda
- clear plastic cups (8 oz)
- clock or watch with a second hand

The soda we drink is made of water, sweeteners, flavorings, and a special gas that makes it bubbly. In this activity, we will use the gas bubbles in soda to have a fun raisin race!

Here's what you do:

1. Fill each plastic cup about $\frac{3}{4}$ full of soda. Place 3 raisins in the soda in each cup. Watch the raisins closely.

2. From each cup, choose the raisin that floated to the top most often. These will be your "prize raisins."

3. Remove the other raisins from the cups. Put one cup aside and use the other cup as your Raisin Challenge Cup!
When you open a soda slowly, the "pssss" you hear is the gas escaping. See if you can find another place where the gasses in soda escaped with too much force!
4. When you say "GO!", you and you partner should place your prize raisins in the challenge cup. The raisin that gets to the surface the most times in two minutes WINS! GOOD LUCK!
GO! GO!

Colorful Candies

This is what you need:

- coated chocolate candy (brown)
- coated peanut butter candy (brown)
- coated fruit jelly candy (brown)
- pencil
- coffee filter (cone type)
- cotton swabs
- water
- paper or plastic cup (at least 7oz.)
- blunt-tip scissors

Some colors we see are really mixtures of other colors. Watch what happens to these yummy brown candies in this activity!

Here's what you do:

1. Cut 3 strips of coffee filter 10 cm long and 3 cm wide. Write the name of each candy on a separate strip as shown.

2. Fill the cup about $\frac{1}{4}$ full with water. Dip one end of the cotton swab into water and use it to wet one side of your candies. Gently rub the candies' wet side onto its filter strip about 2 cm from the end to make a dark dot on the paper.*

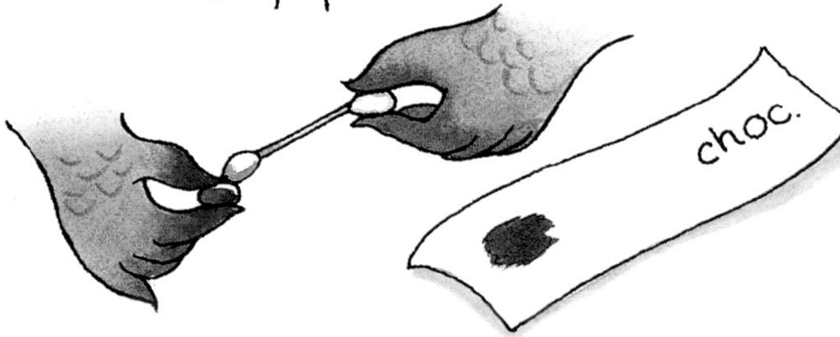

* Do not put the used end of the cotton swab back in the water!

3. Repeat step 2 for the other two candies and paper strips.

4. Carefully place the 3 strips in the cup of water so that only a small portion of the bottom of each strip touches the water.

5. Watch each strip as the water moves up through the dot. What do you see happening? Do you see any colors besides the brown? What colors do you see?

You can check the ingredients listed on the candy wrapper to see if the colors you saw were actually used to color the candies. Now, see if you can find another example of a color mixture being separated somewhere in this book!

Our Sally and Sammy
Like all kinds of sports.
They find lots of science
In sports of all sorts!
The pitchers in baseball
Throw balls that are fast!
The batters must hit them
Before they zoom past!
Snap!

You have to be quick with
Your hands and your eyes,
Or you'll be struck out, Sam,
In only three tries!
But if you would rather
Make up your own game,
Try this one and go for
The game Hall of Fame!
OOPS-SORRY!
You can create a fun game of
your own and discover just how
quick your hands and eyes are
by doing the two activities that
follow.

A Catchy Contest

This is what you need:

- large, lightweight ball

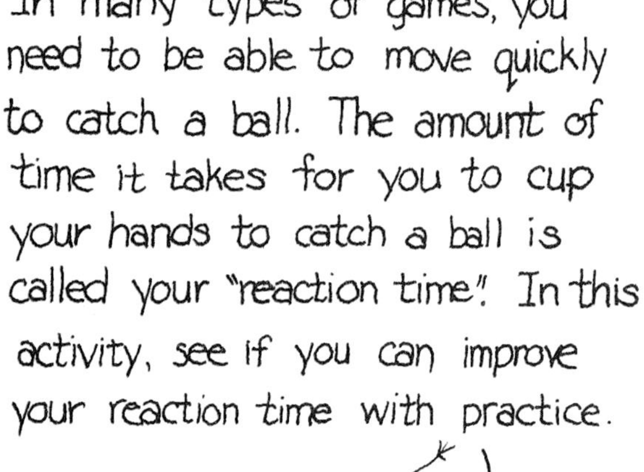

In many types of games, you need to be able to move quickly to catch a ball. The amount of time it takes for you to cup your hands to catch a ball is called your "reaction time". In this activity, see if you can improve your reaction time with practice.

Here's what you do:

1. Stand with your arms at your sides and your hands touching your legs.

2. Have your partner stand behind you holding the ball with both hands. The ball should be held high enough over your head so that you cannot see it.

3. Without telling you when, your partner should drop the ball in front of you.

4. Try to catch the ball. It may not be easy at first, but after a few tries, you will probably get better.

5. If you get really good at catching your partner can make things a bit more challenging by dropping the ball a bit to the right or left.

People who play sports aren't the only ones who need a quick reaction time. See if you can find someone else in this book who has a very short reaction time!

Boing Bat

This is what you need:

- balloon
- blunt-tip scissors
- sturdy plastic cup
- rubber band
- ruler
- tape
- table tennis ball

Many sports use balls made of rubber because of rubber's bounciness. In this activity, you can use the rubber from a balloon to make your own bounce toy!

Here's what you do:

1. Cut the round part of the balloon off and stretch it over the plastic cup. Wrap the rubber band around the rim of the cup so the balloon is held in place.

2. Tape the cup to the ruler so that at least 1 or 2 centimeters is sticking out beyond the cup.

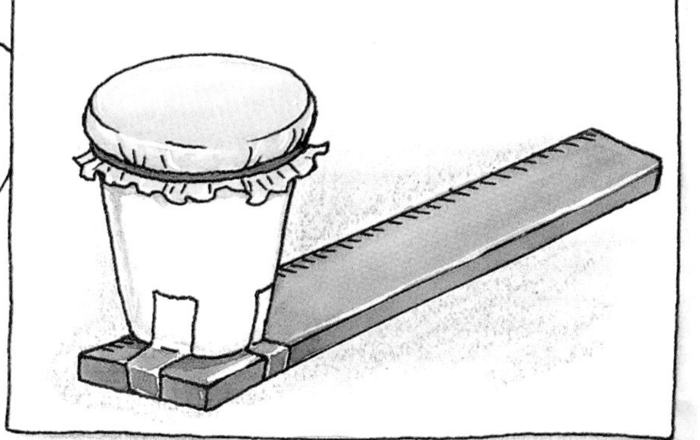

BOING!

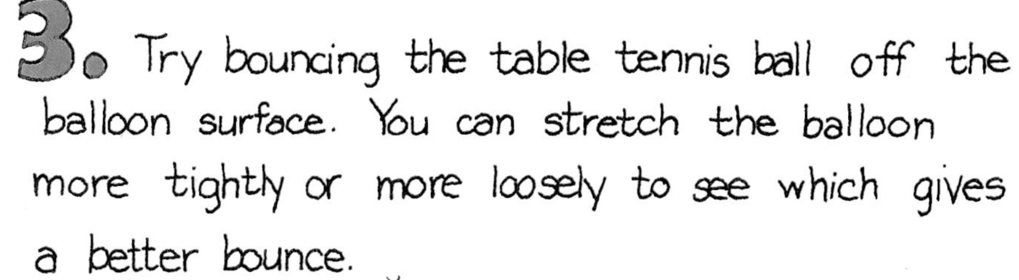

3. Try bouncing the table tennis ball off the balloon surface. You can stretch the balloon more tightly or more loosely to see which gives a better bounce.

It's time to go home now
And help Dad plant seeds.
They're starting a garden—
Let's see where it leads.
I like to plant seeds but
I really must know:
If seeds point the wrong way,
Will my seedlings grow?
However you plant them,
The seeds seem to know
That stems should grow upward
And roots down below!
From plants in our garden
We use every part.
We even use pieces
Of plants for plant art!
SEEDS

LANA
DON'T FORGET
TUES.—
LUNCH WITH
LILLIAN
To learn more about how
seeds grow, and to try
some fun plant art, just
turn the page!

Smart Seeds

This is what you need:

- clear plastic cup (8 oz.)
- paper towels
- bean seeds
- water
- plastic wrap

Here's what you do:

1. Line the inside of the plastic cup with folded paper towels.

2. Crumple a few paper towel sheets and stuff them inside the center of the cup.

3. Add water to the cup to moisten the paper towels.

4. Place 4 bean seeds between the folded paper towel and the cup sides. Position each seed differently.

Now see if you can find another example of a plant that seems to "know" which way it needs to grow.

5. Cover the top of the cup tightly with plastic wrap. Look at your seeds each day. What do you see happening as your seeds sprout?

Awesome Art

This is what you need:

- various plant juices (grape, blueberry, cherry, raspberry)
- flower petals (carnation, violet, hydrangea)
- other plant parts (red cabbage leaves, radish, turnip)
- vinegar
- baking soda
- measuring spoons
- water
- white unlined paper
- cotton swabs
- clear plastic cups
- masking tape
- pencil

Here's what you do:

1. Add 1 teaspoon of baking soda to 3 tablespoons of water in a cup. Label this cup "baking soda".

2. Pour a little vinegar into a cup and label this cup "vinegar".

Plants have been used throughout history for many different purposes. In this activity, you can use plant roots, leaves and juice to make some colorful art!

3. Use the cotton swabs to paint a picture with the different fruit juices. Let your picture dry.

4. Now add more color to your picture by rubbing the flower petals and other plant parts onto the paper.

Now see if you can find another example of a plant part that has left behind some unwanted color!

5. Paint over your picture with the baking soda solution. What happens to the colors in your picture? Now try the vinegar. What do you see happening?

The family's together
At home once again
There can't be more science—
Not here in the den!
Please turn up the volume.
I can't hear the sound.
It's such a good show that
It has me spellbound!
MISS JULIA'S DAY OUT
BIG BOOK of SCIENCE
DICTIONARY

Do you understand how
That sound gets to you?
It vibrates the air and
Your own eardrums, too!
And special effects on
The show that you picked,
Are not what you think 'cause
Your eyes can be tricked!
ouch!
KEEP GOING...
Discover how
sound is made
and how your
eyes can be
tricked!

Shakin' Salt

This is what you need:

- radio
- wax paper
- scissors
- salt

Here's what you do:

1. Cut a square of wax paper about 6 cm x 6 cm.

2. Place the radio on a table and lay the piece of wax paper on the speaker.

wax paper

Sound is made when an object moves back and forth very quickly. These rapid movements are called vibrations. Vibrations travel through the air to our ears. Our ears hear these vibrations as sounds.

See if you can find another example of an object that moves very quickly or vibrates to make sounds!

ILLUSION CONFUSION

This is what you need:

- pencil with eraser
- paper or plastic cup (8 oz.)
- crayons
- several 3x5" index cards
- ruler
- push pin
- blunt-tip scissors
- glue
- tape

Here's what you do:

1. Use your cup to draw a circle on the blank side of an index card and use a ruler to make the wedges. Use red and yellow crayons to color in the wedges.

2. Make a hole in the center of your disk with the pushpin to attach the disk to the eraser end of a pencil.

pushpin
disc
pencil

3. Now spin the pencil quickly back and forth between your hands. What color do you see?

You may think that seeing is believing, but it isn't always! Try this activity and you will find out that sometimes you see things very differently from the way they really are!

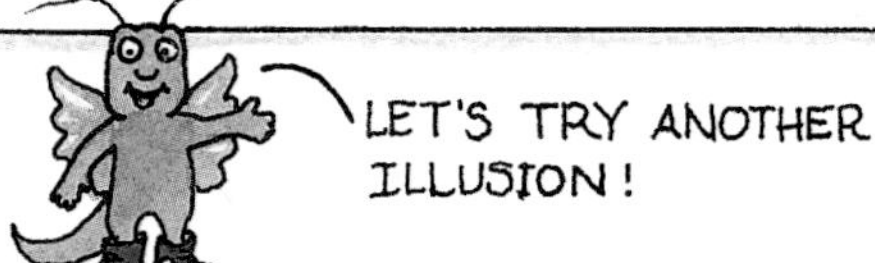

1. Cut an index card into a rectangle the side shown to the right. Find the exact center and draw a line down the middle.

2. Draw a head on one side of the rectangle and a hat on the other side as shown.

3. Fold the rectangle along the center line and tape it to the pencil so that the pictures face outward. Now spin the pencil between your hands. What happens to the hat and the head that you drew?

Optical illusions are things that seem different from the way they really are. See if you can find another example of an optical illusion somewhere in this book!

We now come to bath time—
A nightly routine.
See what they discover
While they're getting clean.

You're right little Sammy
It comes from moist air
It causes small droplets
To form everywhere!
But this water on me
Which came from the tub,
Soaks into my towel
When I start to rub!
Learn more about soaking up water and why mirrors fog up by doing the fun activities on the next two pages!
THIS WAY

Water Wonders

This is what you need:

- 2 clear glass jars or bottles with lids
- water
- ice

Here's what you do:

1. Fill your two jars about halfway with cool water.

In this activity, you can see the cool way the invisible water in the air can make drops we can see!

2. Now place a bunch of ice cubes in one of the jars and put the lids on.

See if you can find another place in the book where water drops form on a cold surface!

3. Watch the outsides of the jars. What do you see happening?

How Wet Does It Get?

This is what you need:

- paper towels
- brown paper bag
- old, white cotton sweat sock
- white, unlined paper
- blue food coloring
- baking pan
- blunt-tip scissors
- pencil
- hanger
- water
- string
- tape

What kinds of materials will soak up water?

Here's what you do:

1. Trace the shark onto the white paper and cut it out. Use the paper shark to cut sharks from the paper bag, the sock and a paper towel.

2. Tape the sharks to the bottom of the hanger.

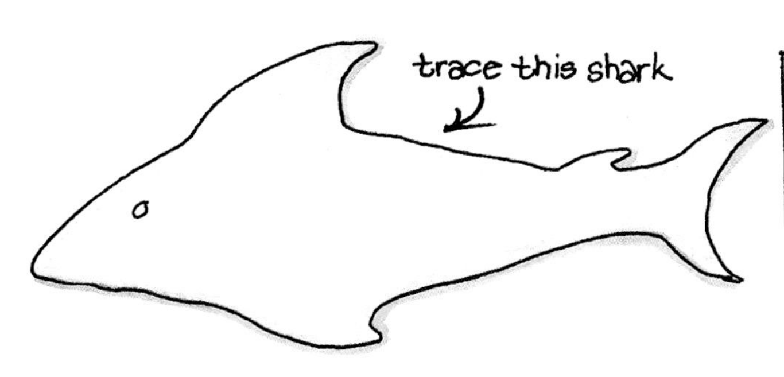

3. Fill the baking pan half full of water and stir in 4 drops of blue food coloring to make an "ocean".

4. Tie a string to the hook of the hanger and suspend the hanger from a cupboard handle so that just the sharks are in the ocean.

5. Watch what happens. Which shark gets completely soaked first?

Which one soaked up the least amount of water?

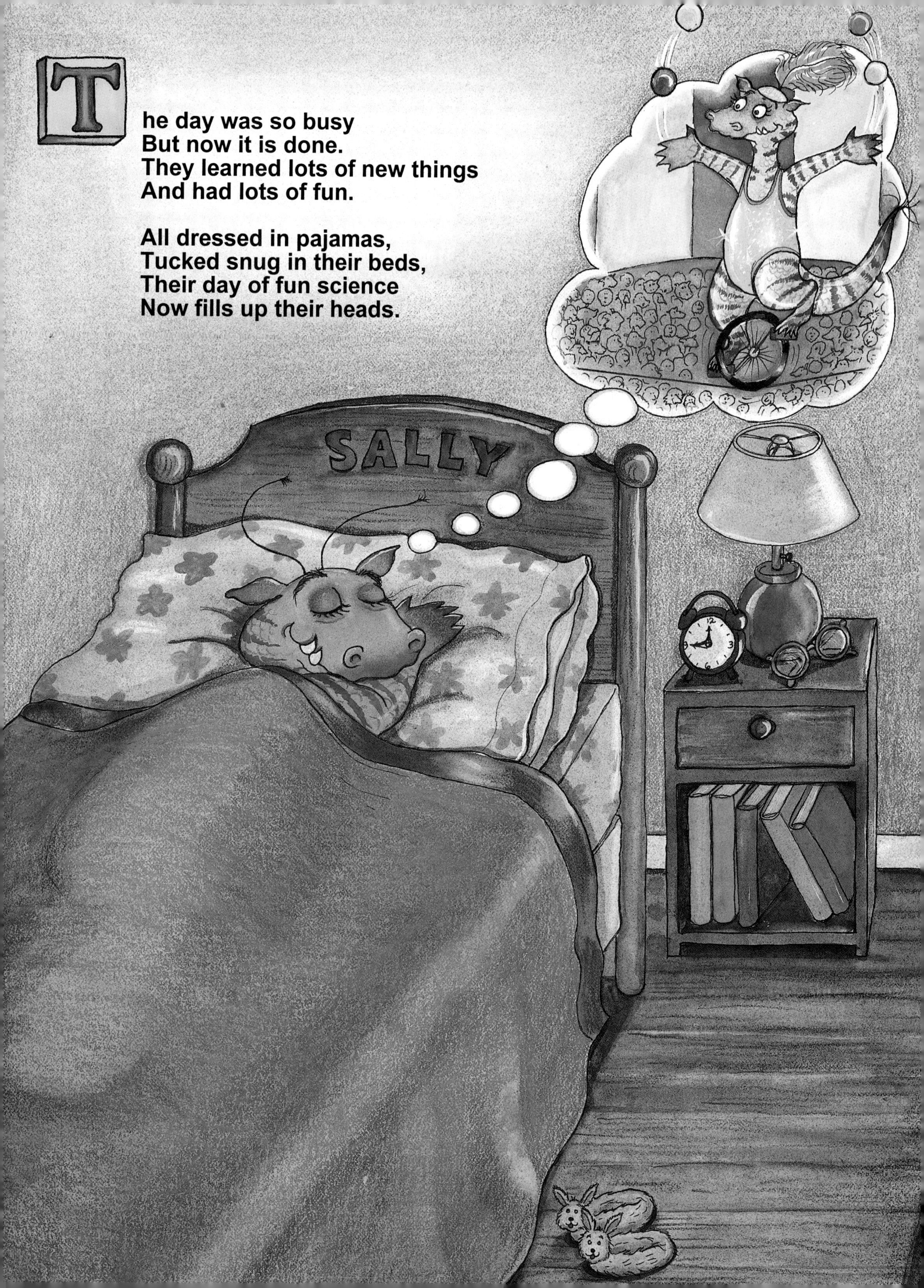

The day was so busy
But now it is done.
They learned lots of new things
And had lots of fun.

All dressed in pajamas,
Tucked snug in their beds,
Their day of fun science
Now fills up their heads.

SAMMY

Science Search Challenge Answers

Static Man
Challenge: Try to find another example of **static electricity**.
Answer: In the den. One of the twins is getting a static shock when touching the dog's nose.

Bouncing Beams
Challenge: Look for another example of **reflection**.
Answer: Walking across the bridge. Sally and Sammy see their reflections in the stream.

Magnet Mystery Map
Challenge: Search for another use for **magnets**.
Answer: In the bedroom. One of the twins is using a magnet to play a fishing game.

The Mighty Dome
Challenge: See if you can find another **dome**.
Answer: At the ballpark. One of the twins is sitting on a batting helmet, which is a strong dome.

Teetering Towers
Challenge: Where is another tall object that uses a **heavy base** to keep it from tipping over?
Answer: At the playground. The basketball pole is supported by its very heavy base.

Bridge Building
Challenge: See if you can find another **bridge**.
Answer: At the friend's house. A bridge is part of the playground equipment.

Balancing Beastie
Challenge: Can you see another example of **balancing**?
Answer: Walking across the bridge. One of the twins is using balance to ride a unicycle.

Surface Sailing
Challenge: Where is another example of something supported by the **surface tension** of water?
Answer: At the garden. Water striders move across the water without falling in because of the water's surface tension.

Racin' Raisins
Challenge: Try to find another example of the effects of **carbon dioxide gas**.
Answer: At the ball park. Soda squirts out of a can because of the carbon dioxide gas inside.

Colorful Candies
Challenge: Where else do you see **colors separating** from each other?
Answer: In the garden. Some colors have run on one of the pieces of laundry.

A Catchy Contest
Challenge: See if you can find someone else who uses fast **reaction time**.
Answer: In the den. The twin's juggling requires quick reaction time.

Boing Bat
Challenge: Locate another toy that works because of its **stretched rubber**.
Answer: At the friend's house. One of the twins is bouncing a big rubber ball.

Smart Seeds
Challenge: Can you find another example of plants seeming to know **which way to grow**?
Answer: In the bedroom. The plant on the window sill is bending and growing toward the light.

Art Adventures
Challenge: Try to find another example of a **color that comes from a plant part**.
Answer: In the bathroom. Some clothes on the floor have green grass stains.

Dancing Salt
Challenge: Look for someone making **sounds from vibrations**.
Answer: In the kitchen. The guitar strings the twin is playing vibrate to make noise.

Illusion Confusion
Challenge: Where is there another example of an **optical illusion**?
Answer: In the bathroom. The 3-D glasses the twin is wearing to read a 3-D comic book create an optical illusion.

How Wet Does It Get?
Challenge: Find another example of a **material soaking up wetness**.
Answer: At the playground. One of the twins is playing basketball and wearing wrist bands and a headband made from a material that soaks up sweat.

Water Wonders
Challenge: Where else do you see **water droplets forming on a cold surface**?
Answer: In the kitchen. Water droplets form on the milk container when air condenses on its cold outer surface.

What's the Science Behind the Fun?

Hands-on Activity Explanations

1. Static Man

When a balloon is rubbed on hair, extra electrons (negative charges) from the hair end up on the balloon. When the balloon is brought near the pieces of thread, the electrons from the balloon repel some electrons from the tips of the strings, leaving extra protons (positive charges) in the string. The electrons on the balloon and the protons on the strings then attract each other.

2. Bouncing Beams

When light strikes an object, it may pass through the object, be absorbed by the object, or bounce off the object. A mirror, which has a very smooth surface, causes light to bounce off of or be "reflected" from its surface. Reflected light can be aimed, because the angle at which the light hits the mirror will always equal the angle at which the light bounces off.

3. Magnet Mystery Map

Magnetism is caused by tiny electric currents produced by moving electrons in atoms. In magnetic materials, such as iron, nickel, and cobalt, the electric currents act together in a special way to make the materials magnetic. The materials that can be made into magnets are the same ones that can be attracted by a magnet. Aluminum, wood, and plastic are not magnetic.

4. The Mighty Dome

A fragile material such as eggshell can support a surprising amount of weight if it has the right shape. A dome or an arch shape spreads out the forces pressing down on the material, enabling it to support a great amount of weight.

5. Teetering Towers

Adding pennies to the bottom cup of the paper cup tower makes the tower more stable and better able to withstand the "wind storm." Just like a real building, the taller the paper cup tower, the more massive the base must be. As real skyscrapers get taller, their foundations need to be deeper and more massive.

6. Bridge Building
Bridges made with different designs will support different numbers of pennies. Generally, if the sides of the bridge are bent upward, the bridge will be stiffer and can support more weight. The accordion design takes advantage of this by featuring many vertical bends. An arch under the bridge may also do a good job. Like the eggshell domes in "The Mighty Dome" activity, the rounded shape of the arch spreads out the weight of the pennies, allowing the bridge to support more weight.

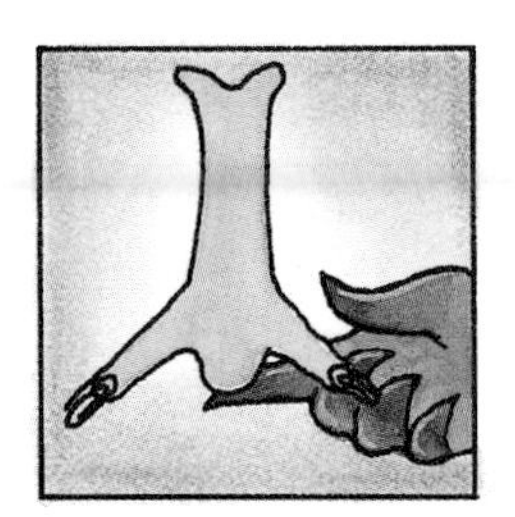

7. Balancing Beastie
How an object balances depends on the location of its **center of gravity.** The center of gravity is the point around which the weight of the object is equally distributed. In the case of "Balancing Beastie," the heavy paper clips put his center of gravity between his arms. When you flip him upside down, with his head on your finger, his center of gravity is right below your finger. Because the forces are pulling down evenly on both sides of your finger, he can balance on his head.

8. Surface Sailing
The index card boat appears to float on the water's surface, but it is actually resting on the top layer of water molecules, which acts like a strong but flexible "skin "stretched across the surface of the water. Adding a drop of detergent breaks this skin behind the boat. The rest of the skin tears away from this broken area, carrying the boat with it.

9. Racin' Raisins
When placed in a cup of soda, raisins move up and down because of the carbon dioxide gas in the soda. The bubbles of gas attach to the raisins like little life preservers and give the raisins enough buoyancy to float to the surface. At the surface, the gas bubbles burst, causing the raisins to sink back to the bottom, where they pick up more gas bubbles and repeat the process.

10. Colorful Candies
Just as an artist mixes colors for a painting, candy makers mix different color dyes to use in candy coatings. The different colors are made up of different **pigment** molecules. The colors separate because the different pigment molecules move up the coffee filter strip at different rates.

11. A Catchy Contest
To catch the falling ball, not only do your hands and eyes need to work together, but so does your brain. When the ball comes into view, a message is sent from your eyes to your brain. Your brain then sends a message to your hands to catch the ball. The speed and accuracy of this eye-brain-hand system can improve with practice.

12. Boing Bat
The balloon used to make the boing bat is made of rubber. Because of the structure of the molecules that make up rubber, when rubber is stretched and then let go, it returns quickly to its original shape. This quality makes rubber an excellent surface on which to bounce a ball and an excellent material for making a bouncing ball.

13. Art Adventures
Plants contain colorful chemicals called **pigments.** Sometimes, when another chemical, such as an **acid** like vinegar or a **base** like baking soda, is added to these pigments, the pigment changes color. On the basis of the color change, scientists can learn something about the chemical that was added to the pigment.

14. Smart Seeds
Whether a seed is planted pointing up, down, or sideways, the stem always grows up and the root always grows down. The tips of the root and stem contain special structures and chemicals that control the growth of their cells. Slight differences in the cells' growth rate on one side of the root or stem cause the plant to bend and to grow in the proper direction.

15. Dancing Salt
Sound is produced from vibrating objects, which make air molecules move back and forth. The air molecules can hit other objects, like your ear drum, and make it vibrate, too. In this activity, the sound from the radio vibrates the wax paper, causing the salt to bounce on its surface. Changing the volume or kind of music changes the vibrations, making the salt move in different ways.

16. Illusion Confusion
The illusion in this activity is a result of our eyes retaining an image of an object they have just seen for a fraction of a second after it is gone. The eyes retain one image while looking at another and the brain puts the two together. That is why the color wheel looks orange (the eyes see the red and yellow "on top of one another") and why the hat looks like it is on the man's head (the brain is putting the separate images of the man and the hat together).

17. How Wet Does It Get?
Materials such as cloth and paper soak up water mainly for two reasons. The first is that the tiny fibers that make up the material attract, rather than repel, the water molecules. Second, these materials have many tiny narrow spaces that can fill up with water, so the water can "climb" up the material.

18. Water Wonders
There are water molecules in the air all around us. This water in the air is called **water vapor.** When water vapor cools, the water molecules begin to stick together, or **condense**, to form tiny invisible droplets of liquid water. When enough of these tiny droplets collect on cold surfaces, like the outside of a cold jar, they form drops that are large enough to see.

About the Book

The Authors

James Kessler currently is manager of the K-to-8 Science Office in the Education Division of the American Chemical Society (ACS). He has been the editor of ACS's *WonderScience* magazine, an elementary school hands-on science activity magazine, since 1989. James also oversees the production and dissemination of other science educational materials for students and teachers in grades K-to-8 and conducts hands-on science workshops for pre- and in-service teachers. James received a B.A. in philosophy from Columbia University, a J.D. from Boston University School of Law, and a B.S. in science education from the University of Maryland. He has taught biology and physical science in the Washington, D.C. area and in São Paulo, Brazil. James is co-author of *Apples, Bubbles, and Crystals.*

Andrea Bennett, an experienced science teacher, develops science teaching resources for ACS's K-to-8 Science Office. Andrea currently conducts hands-on science workshops for children with the Smithsonian's Young Associates program. She holds bachelors degrees in both biology and science education from the University of Maryland. Andrea is co-author of *Apples, Bubbles, and Crystals.*

The Illustrator

Melody Sarecky, a Corcoran School of Art graduate, has been illustrating for children's publications for 15 years, including *WORLD* Magazine, *HSUS News, Ranger Rick, Science Scope,* and *Weekly Reader.* Books that she has illustrated include *Serious Fun; Dragon's Lunch; One Dad, Two Dads, Brown Dad, Blue Dads;* and *Apples, Bubbles, and Crystals.*

The American Chemical Society

The **American Chemical Society (ACS),** a non-profit professional organization of chemists, chemical engineers, and teachers, publishes the world's most prominent chemistry journals and serves as the leading information resource on chemical science and technology. The Society also develops science curricula, supports teacher training workshops, and provides science information for students of all ages and levels.

The **ACS Education Division** supports the development and implementation of programs that bring the wonder, excitement, opportunities, and challenges of modern chemistry to students of all ages. For more information about Education Division programs write to:
Education Division, American Chemical Society,
1155 16th Street, N.W.,
Washington, DC 20036
or call (202) 452-2113.